Cancer Cachexia

A Comprehensive Guide to Cachexia, Clinical Implications, Diagnosis, Treatment Strategies and Related Conditions

Isabella White

Copyright © 2024 by Isabella White.

The publisher reserves all rights. Without prior written permission, no part of this publication may be reproduced, distributed, or transmitted in any form or by any means, including photocopying, recording, or other electronic or mechanical methods. Under copyright law, limited quotations may be used for non-commercial purposes and critical reviews.

Disclaimer: The information in this book is based on the author's research, opinions, and experiences. It is not intended to replace professional medical advice or treatment. The reader should regularly consult a physician for any health issues and always seek the advice of a physician before modifying diet, supplement, or exercise regimens. The author and publisher shall have neither liability nor responsibility to any person or entity concerning any loss or damage related to the information contained in this book. The information provided is general and may not apply to every individual. Any reliance on the information contained herein is solely at the reader's risk.

Table of Contents

Introduction _____ 5

Chapter 1
Understanding Cachexia _____ 8
    Definition and Description of Cancer Cachexia ____ 8
    Differentiating Cachexia from Other Conditions __ 11
    The Prevalence and Impact of Cancer Cachexia __ 13
    Pathophysiology and Mechanisms Underlying
    Cancer Cachexia _____ 16
    Signs, Symptoms, and Clinical Presentation _____ 19
    Classification Systems and Staging _____ 22

Chapter 2
Diagnosing Cachexia _____ 26
    Diagnostic Criteria and Consensus Definitions ___ 26
    Methods of Assessing Weight Loss and Body
    Composition _____ 30
    Potential Biomarkers and Lab Tests to Support
    Diagnosing Cancer Cachexia _____ 33
    Assessment Tools and Questionnaires _____ 36
    Challenges in Diagnosis of Cancer Cachexia _____ 39

Chapter 3
Clinical Implications and Prognosis _____ 42
    The Impact of Cachexia on Clinical Outcomes ___ 42
    Cachexia Staging and Expected Prognosis _____ 45
    Effects of Cancer Cachexia on Quality of Life ____ 48
    Health Economics and Healthcare Utilization

    Related to Cancer Cachexia _____ 51
    Ethical Considerations Related to Cancer Cachexia _____ 55

Chapter 4
Prevention and Management _____ 58
    The Goals of Cancer Cachexia Treatment _____ 58
    Pharmacologic Therapies _____ 61
    Nutritional Interventions _____ 63
    The Role of Exercise and Physical Activities _____ 66
    Behavioral Strategies _____ 69
    Multimodal Approaches _____ 71
    Evaluating the Treatment Response _____ 74
    Challenges in the Treatment of Cancer Cachexia _____ 77

Chapter 5
Special Considerations _____ 80
    Cachexia in Association With Other Chronic Conditions Beyond Cancer _____ 80
    Unique Factors Related to Cachexia in Certain Types of Cancer _____ 83
    Important Considerations for Cancer Cachexia Prevention and Management Across the Cancer Trajectory _____ 86
    Special Considerations for Cancer Cachexia Prevention and Management In Vulnerable Populations _____ 88
    Cachexia Care Considerations in Palliative and End-of-Life Settings _____ 91

Conclusion _____ 94

Introduction

Cancer cachexia is a complex and debilitating condition that impacts a significant portion of cancer patients, profoundly impacting quality of life and clinical outcomes. However, it remains underdiagnosed and undertreated.

This book aims to provide a comprehensive guide to cachexia to support improved recognition, prevention, and management of this condition. Within this book, readers will gain a thorough understanding of the mechanisms and clinical presentation of cachexia.

The intricate interplay of the disease process, inflammation, metabolic disturbances, and reduced food intake is investigated, providing context for the various manifestations of involuntary weight loss, muscle and fat mass loss, anorexia, inflammation, and functional impairment. Guidance is provided to strengthen assessment skills and highlight the subtle signs of early cachexia.

Current cachexia screening and diagnosis best practices are reviewed in detail, covering diagnostic criteria, assessment tools, and potential biomarkers under investigation. Challenges and controversies in measurement and staging are discussed. The importance of early identification to enable timely supportive care and cachexia management is emphasized throughout.

Nutritional, exercise, pharmacologic, and psychosocial interventions are described to equip readers with practical cachexia management strategies. Both conventional approaches and promising emerging therapies are included. Guidance on developing individualized multimodal care plans is provided, focusing on optimizing patient outcomes and quality of life.

The impact of cachexia across the cancer journey is explored, highlighting clinical implications at diagnosis, during active treatment, in advanced disease, at end-of-life, and in cancer survivors. Cachexia's effects on treatment tolerance, risks, efficacy, and sequelae are examined. Insights are provided into tailoring care for patients with cachexia to improve outcomes.

Related syndromes, including sarcopenia, anorexia, and malnutrition, are differentiated and demystified. Assessment, diagnosis, and management strategies are compared and contrasted with cachexia throughout the book to support identification and integrated care.

Based on the most recent research and practical insights, this in-depth resource aims to advance the understanding of cancer cachexia and equip healthcare professionals to provide optimal care to patients experiencing this condition.

Cancer care professionals who are dedicated to improving the lives of people affected by cachexia, such as oncology nurses, doctors, dietitians, physical therapists, palliative care specialists, and others, should read this useful handbook.

Chapter 1

Understanding Cachexia

Definition and Description of Cancer Cachexia

Cancer Cachexia is a multifactorial syndrome defined by the involuntary loss of skeletal muscle mass that cannot be fully reversed through conventional nutritional support. It results from an unhealthy protein and energy balance driven by variable combinations of reduced food intake and abnormal metabolism.

Cachexia is clinically manifested through weight loss, muscle wasting, loss of appetite, inflammation, insulin resistance, and functional impairment.

At the cellular level, cachexia reflects an imbalance between widespread catabolic processes breaking down tissues and disrupted anabolic processes required to maintain muscle. Pro-inflammatory cytokines, activated by the tumor and host responses, trigger heightened protein degradation and lipolysis. This contributes to lean tissue depletion and fat loss. Skeletal muscle represents the main cachectic target, resulting in reduced strength, fatigue, mobility challenges, and a diminished quality of life.

Cachexia emerges as a consequence of tumor presence and mutual interactions between the tumor and host. Pro-cachectic factors produced by certain cancers directly induce metabolic abnormalities and drive the syndrome's progression. Even in the absence of these factors, cachexia may still develop, accelerated by the inflammation arising from the host immune response to malignancy.

Cachexia represents a continuum, ranging from pre-cachexia to refractory cachexia. Pre-cachexia describes early clinical and metabolic signs, including anorexia and impaired glucose tolerance, before substantial involuntary weight loss. Refractory cachexia represents end-stage progressive muscle loss that is unresponsive to treatment. Staging supports prognosis and guides management.

Cachexia prevalence varies substantially based on cancer type, location, and stage. The highest rates are seen in cancers of the lung, pancreas, and gastrointestinal tract. Up to 80% of advanced cancer patients develop cachexia, with 20% ultimately dying as a direct consequence. Certain populations, such as elderly, sedentary, or already malnourished patients, face heightened cachexia risks.

If unaddressed, cachexia contributes significantly to morbidity and mortality in cancer patients. It worsens outcomes and quality of life across the care continuum. Screening and prompt interventions focused on its underlying mechanisms can help mitigate cachexia's impact and support patients through their cancer journey. Increased recognition of this complex, common, and costly syndrome is critical to improving patient care.

Differentiating Cachexia from Other Conditions

Cachexia shares features with several related syndromes, including sarcopenia, malnutrition, and anorexia. However, important distinctions exist.

Sarcopenia refers specifically to the loss of skeletal muscle mass and strength. Cachexia includes muscle wasting but also encapsulates the depletion of fat stores and other systemic manifestations. Sarcopenia may develop in cachexia but can arise independently from disuse, aging, or disease. Different mechanisms drive primary sarcopenia versus cachexia-induced muscle loss.

Malnutrition denotes inadequate caloric and nutrient intake. It may result from reduced appetite, digestive issues, or food access. Malnutrition contributes to the protein and energy deficit of cachexia but does not fully account for the syndrome's metabolic abnormalities and tissue breakdown. Patients may be cachectic despite sufficient caloric intake due to hypermetabolism. Alternatively, malnourishment alone does not necessarily induce cachexia without the inflammation and catabolism triggered by malignancy.

Anorexia represents a loss of appetite and an inability to maintain caloric intake. Decreased oral food intake is frequently associated with cachexia and contributes to weight loss. However, cachexia-related anorexia arises for complex metabolic reasons beyond behavior. Patients may eat sufficiently but remain cachectic due to upregulated energy expenditure and abnormal protein turnover. Equally, simple disinterest in food without underlying metabolic dysfunction does not automatically signal cachexia.

These syndromes interact significantly. Anorexia worsens the energy deficit in cachexia. Sarcopenia and malnutrition exacerbate wasting. Cachexia is often associated with underlying sarcopenia and malnutrition. Cachexia-associated inflammation and catabolism can independently drive anorexia.

However, cachexia involves unique characteristics beyond these associated conditions. Isolating cachexia centers on the identification of involuntary weight loss plus underlying metabolic change. This guides targeted nutrition and pharmacology interventions to address its distinct mechanisms.

Careful history and assessment help differentiate cachexia from primary malnutrition, sarcopenia, or anorexia and direct appropriate management. Evaluation of caloric intake, fat and lean mass measurement, diagnostic bloodwork, quality of life metrics, and patient-reported appetite changes support clinicians in delineating overlapping syndromes.

A multidisciplinary approach combining perspectives from medical oncology, nutrition, and rehabilitation services facilitates accurate identification of cachexia where present. Distinguishing cachexia from other wasting conditions is critical for prognosis and enabling optimal treatment across the cancer care continuum.

The Prevalence and Impact of Cancer Cachexia

Cachexia is a frequent complication of cancer, occurring in over half of all patients with malignant disease. However, prevalence varies substantially depending on cancer type, location, stage, and patient factors. The highest rates of cachexia are observed in cancers of the gastrointestinal tract, lung, and pancreas.

Among pancreatic cancer patients, more than 80% experience cachexia during their disease course. Over 60% of those with gastric or esophageal malignancies also become cachectic. Approximately one-third of people with lung cancer are cachectic at diagnosis, rising to over 60% in end-stage disease. Rates also remain high in lymphoma, prostate, colorectal, and other cancers.

Advanced cancer correlates with increased cachexia prevalence. Up to 20% of cancer patients are cachectic at initial diagnosis, climbing to 50–80% in those with end-stage or refractory disease. Palliative populations frequently bear the deepest burden of cachexia and its effects on quality of life.

Patient variables likewise influence susceptibility. Cachexia most commonly develops in those over the age of 65. Men tend to be more predisposed than women. Concurrent sarcopenia, anorexia, malnutrition, and inflammation increase the risk. Genetic polymorphisms related to cytokine expression may also make certain patients more cachexia-prone.

Awareness of cancer-specific and patient-related risk factors supports clinicians in focusing screening and preventive efforts where they are likely to yield the greatest impact. Unfortunately, cachexia remains underdiagnosed despite its frequency, often not recognized until physical manifestations become overt. Early identification is crucial to mitigating cachexia's profound consequences.

Unmanaged cachexia contributes significantly to morbidity and mortality across cancer types. It is estimated to be directly implicated in up to 20% of cancer deaths. Muscle wasting impairs physical function, reduces treatment tolerability, and diminishes quality of life. Cachexia also drives treatment resistance, increases the toxicity of chemotherapy, and reduces survival duration. Addressing cachexia is critical to supporting patients through the cancer journey from diagnosis to palliation.

Increased recognition and understanding of cachexia have gradually improved over recent decades. However, ongoing awareness and education are essential to drive earlier identification, informed prevention, and integrated cachexia-focused care. Routine screening, improved diagnostics, patient education, and the implementation of evidence-based interventions will be instrumental in reducing the burden of this debilitating syndrome.

Pathophysiology and Mechanisms Underlying Cancer Cachexia

Cachexia arises from a complex interplay of metabolic disturbances, cytokine-driven inflammation, and reduced nutritional intake. While multiple mechanisms contribute, chronic systemic inflammation appears central to tipping the balance towards uncontrolled tissue catabolism.

Pro-inflammatory cytokines, including TNF-alpha, IL-1, IL-6, and interferon-gamma, are critical drivers.

These are produced by immune cells in response to the presence of the tumor and by the tumor itself. Cytokines activate intracellular pathways that increase muscle protein breakdown and suppress synthesis, resulting in muscle wasting. They also stimulate lipolysis and adipose tissue loss while inhibiting lipogenesis.

Cytokines drive widespread metabolic dysfunction. They impair carbohydrate and protein metabolism, increase resting energy expenditure, and induce insulin resistance. Mitochondrial abnormalities further dysregulate metabolism and energy balance. Highlight catabolism outpaces the anabolic processes required to maintain lean tissues and adipose stores.

In addition to stimulating proteolysis, cytokines suppress appetite and reduce oral intake. Tumor byproducts may directly induce anorexia. Reduced ingestion exacerbates weight loss and nutrient deficiencies, accelerating muscle wasting in a positive feedback loop.

Muscle represents the primary target of cachexia catabolism. Depletion of contractile proteins, including actin and myosin, reduces muscle fiber diameter, strength, and function. However, all organs can be impacted. Cardiac atrophy impairs function. Bone wasting increases fracture risks. Hepatic dysfunction disrupts metabolism. Neural changes may underlie anorexia and fatigue.

There is considerable inter-individual variability in pathways activated based on cancer type and host factors. Pancreatic tumors often overexpress lipid mobilizing factors, driving aggressive fat-store catabolism. Certain cancers produce proteolysis-inducing factors to stimulate muscle wasting. Host genetics influence cytokine responses. This heterogeneity complicates efforts to identify universal cachexia biomarkers and treatments.

While inflammation enables cachexia, muscle breakdown ultimately depends on activating the ubiquitin-proteasome system, which marks proteins for destruction. Ongoing research seeks to clarify regulators of this final common pathway as therapeutic targets.

An increased understanding of cachexia's pathophysiology highlights that it is beyond mere starvation but rather a complex metabolic disorder requiring tailored interventions. Combining anti-inflammatory, anabolic, and nutritional therapies continues to be an integrated treatment approach.

Signs, Symptoms, and Clinical Presentation

Cachexia manifests through a range of physical, functional, and psychological symptoms stemming from the underlying loss of skeletal muscle and fat. Clinical presentations vary based on cancer type, cachexia stage, and individuals. Key signs center on involuntary weight loss, frequently accompanied by anorexia, inflammation, and metabolic disturbances.

Unintentional weight loss of greater than 5% body mass over six months is a cardinal diagnostic feature. A more profound depletion of over 10% may be observed as cachexia progresses. Wasting of skeletal muscle and adipose tissues contributes to weight loss; fluid shifts between compartments may also occur.

Declining muscle mass directly reduces strength and physical function.

Anorexia and early satiety are common, stemming from cytokine-driven appetite suppression and tumor byproducts. Reduced oral intake exacerbates weight loss and augments muscle wasting. Taste changes may also develop. Resultant malnutrition and nutrient deficiencies further affect tissue integrity and patient health.

Systemic inflammation manifests through fever, elevated inflammatory markers like CRP, anemia, altered glycemic control, insulin resistance, and hypoalbuminemia. Plasma proteins decline as liver synthesis slows and capillary permeability increases. Chronic inflammation both promotes and results from cachexia.

Beyond weight changes, close examination may reveal a loss of muscle bulk. Temporal or deltoid muscles are often visibly wasted. Expanded intermuscular spaces emerge as tissues deplete. Reduced skin turgor from fat loss gives rise to loose skin folds. Ascites and peripheral edema can develop as albumin declines.

As muscle catabolism advances, patients experience weakness, fatigue, reduced endurance, and impaired mobility. Physical function deteriorates, reflected in poor performance on grip strength, get-up-and-go time, and 6-minute walk assessments. Balance, posture, and respiratory function may also decline.

Mood changes, including depression and anxiety, frequently accompany cachexia, stemming from pro-inflammatory cytokines, tumor byproducts, and the quality of life impacts of muscle wasting. Emotional distress and social withdrawal may result.

Careful history, physical examination, and body composition analysis allow clinicians to detect early signs of cachexia before overt wasting, so timely interventions can be implemented to slow its progression and preserve quality of life. Patients may not self-report their initial weight loss. Awareness of various clinical manifestations supports prompt identification and management.

Classification Systems and Staging

Several classification systems have been developed to define and stage the progression of cancer cachexia. These aim to standardize terminology, guide prognosis, and direct management interventions at appropriate points along the cachexia continuum.

An international consensus group proposed one of the most utilized criteria in 2011. This delineates cachexia as weight loss greater than 5% over six months or BMI less than 20 plus any three of decreased muscle strength, fatigue, anorexia, low fat-free mass index, or abnormal biochemistry, including increased inflammatory markers, anemia, or low serum albumin. These consensus criteria classify patients as follows:

1. **Pre-cachexia:** Early clinical and metabolic signs (anorexia, impaired glucose tolerance) before substantial weight loss.
2. **Cachexia:** Involuntary weight loss >5% or BMI <20 plus clinical criteria.
3. **Refractory cachexia:** End-stage cachexia unresponsive to treatment.

The criteria facilitate identifying cachexia before severe wasting develops, so supportive care can aim to slow further progression. Limitations include reliance on weight loss thresholds when metrics of muscle mass may better reflect early cachexia. Emerging research seeks to refine definitions and diagnostic cutoffs.

Other classification approaches incorporate more stages. A four-stage system proposes:

- Pre-cachexia
- Cachexia: Mild (early)
- Cachexia: Moderate
- Cachexia: Severe (refractory)

This supports differentiating the depth of cachexia to guide management. However, it needs more definitive thresholds delineating stages.

Efforts to add muscle mass quantification have yielded staging systems integrating the degree of depletion. One approach defines:

- Pre-cachexia
- Minor cachexia (weight loss <10%)

- Moderate cachexia (10–15% weight loss)
- Severe cachexia (>15% weight loss)

Muscle mass measurement via CT, MRI, or DEXA supports a more precise classification, but availability limits adoption. Ongoing research pursues optimal methods to stage cachexia that balance feasibility with precision to guide prognosis and treatment.

Each universally accepted classification system currently exists. In practice, clinicians utilize elements of different criteria while incorporating additional factors like tumor stage, serum biomarkers, and quality of life impacts to characterize cachexia severity. Staging helps convey the prognosis and stratify management strategies, aiming to stabilize cachexia where possible or slow further progression. More work is needed to optimize standardized methods for classifying this multidimensional syndrome.

Clear communication of cachexia severity through consistent terminology and classification enables the care team to set appropriate goals and select therapies matched to where patients fall on the spectrum from pre-cachexia to refractory cachexia. A shared understanding of the stage supports aligning expectations and priorities for cachexia-focused care.

Chapter 2

Diagnosing Cachexia

Diagnostic Criteria and Consensus Definitions

There is no single universally accepted set of diagnostic criteria for cancer cachexia. However, international expert groups have proposed consensus definitions to standardize identification and guide clinical practice.

In 2011, a panel of the Society of Cachexia and Wasting Disorders established diagnostic criteria that have become widely utilized in research and clinical care. These define cachexia as:

- Weight loss >5% over the past six months (in the absence of simple starvation) **OR**
- BMI <20

Plus any three of:

- Decreased muscle strength,
- Fatigue,
- Anorexia,
- Low fat-free mass index,
- Abnormal biochemistry (anemia, elevated inflammatory markers, low albumin).

These criteria allow for the diagnosis of cachexia at earlier stages before severe muscle wasting develops. However, some limitations exist. BMI lacks sensitivity, as a low BMI may reflect pre-existing sarcopenia, not recent cachexia. The threshold weight loss required overlooks cachexia without marked depletion. Emerging research suggests lower cutoffs (2–5%) may better identify early cachexia.

In 2018, an update by the European Palliative Care Research Collaborative refined diagnostic criteria. Their definition requires:

- Weight loss >5% over the past six months (or BMI <20)
- Plus any three of:
    - Decreased muscle strength
    - Fatigue
    - Anorexia
    - Low fat-free mass index
    - Abnormal biochemistry **OR**
- Low appendicular skeletal muscle index consistent with sarcopenia

These updated criteria demonstrate increased sensitivity without compromising specificity based on validation studies. They incorporate an appendicular muscle index assessment to identify cachexia in patients without pronounced total weight loss.

Both consensus definitions provide a foundation for standardized diagnosis, staging, and enrollment in clinical trials. However, work is still being done to improve cutoffs as new information about early biomarkers and the best ways to spot cachexia before it leads to extensive wasting. Opportunities remain to improve criteria based on muscle quantification, inflammation markers, and metabolites.

In practice, cachexia diagnosis integrates consensus criteria with clinical judgment based on symptom duration, expected weight trajectories, tumor behavior, and patient reports. Diagnosis can be challenging during active anti-cancer treatment when weight fluctuations from side effects complicate the assessment. A multidisciplinary, multiparameter approach combining perspectives from medical oncology, nutrition services, physical therapy, and nursing care facilitates accurate identification.

While work continues to optimize definitions, adopting available consensus criteria enables consistent identification of cachexia to guide prognosis, staging, and supportive care planning. Efforts to standardize diagnosis promote improved recognition and response to this undertreated syndrome.

Methods of Assessing Weight Loss and Body Composition

Accurately tracking weight loss and body composition changes are essential for cachexia diagnosis and monitoring intervention responses. Several methods exist with varying feasibility, precision, and availability.

Weight is most easily assessed via routine scale measurements during clinical encounters. Any involuntary weight loss exceeding 5% over six months raises concern for cachexia, requiring further evaluation. More frequent weight checks allow detailed trend analysis, although fluctuations from dehydration or fluid retention can complicate interpretation.

Body mass index (BMI) adjustments for height provide additional context, although low BMI alone does not confirm cachexia if reduced weight reflects longstanding small stature. BMI lacks sensitivity for early losses of muscle masked by fluid retention or remaining adipose stores.

Skin fold thickness measured at standard anatomical points offers a simple proxy for subcutaneous fat stores. Reduced skin folds suggest fat loss consistent with cachexia. However, this method is operator-dependent and does not quantify the depletion of visceral or intramuscular fat compartments.

Waist circumference measurement can screen for reduced truncal muscle and fat mass. However, it does not distinguish the contributions of specific tissues. Changes may lag behind systemic cachexia progression.

Advanced techniques provide direct visualization and quantification of soft tissues. Computed tomography (CT) and magnetic resonance imaging (MRI) permit precise measurement of muscle cross-sectional area, muscle radiodensity, and adipose stores. However, radiation exposure and the availability of CT and MRI limits frequent use.

Dual-energy X-ray absorptiometry (DEXA) is an easily accessible option for noninvasively assessing fat mass, lean mass, and bone density. It exposes patients to minimal radiation and facilitates safe, serial monitoring of body composition changes.

Bioelectrical impedance analysis (BIA) estimates body composition by measuring resistance to a low electrical current. Changes in fluid distribution from cachexia can skew results. BIA remains convenient for screening and monitoring if limitations are recognized.

There is no single approach that is the gold standard. Combining anthropometry, imaging, BIA, and other data facilitates a comprehensive analysis of tissue catabolism. Even basic weight checks provide value for screening and trend monitoring if more advanced body composition testing proves unavailable. The key is consistently using the same techniques over time for individual patients.

Careful attention to weight patterns, fat loss, and muscle depletion from cachexia enables quantification of the extent of wasting to guide staging, prognosis, and evaluation of therapeutic efficacy. Serial monitoring finds progressive cachexia so that it can be treated quickly to stop the ongoing catabolism and the problems it causes.

Potential Biomarkers and Lab Tests to Support Diagnosing Cancer Cachexia

Cachexia arises from complex molecular pathways and metabolic disturbances. Ongoing research seeks to identify biomarkers that could facilitate the diagnosis and tracking of cachexia progression or treatment response. While no definitive laboratory tests are available at the moment, several show potential promise.

Inflammatory markers offer readily available candidates to reflect cachexia-associated inflammation. C-reactive protein (CRP), interleukin-6 (IL-6), and interleukin-1 (IL-1) often rise with cachexia. However, inflammation alone is nonspecific. Elevated CRP levels accompany many cancer-related conditions beyond cachexia. The degree of change does not reliably indicate cachexia severity.

Emerging inflammatory biomarkers like osteoprotegerin show links to cachexia but require more prospective studies. Recent work proposes combining CRP or IL-6 with serum albumin as a cachexia inflammation panel, but validation is needed.

Insulin resistance and glycosylated hemoglobin levels frequently increase with cachexia, although not universally. These markers could strengthen diagnosis when incorporated into multidimensional criteria but should not be used in isolation. As anemia increases because of inflammation and cytokine release caused by cachexia, hemoglobin, hematocrit, and red blood cell indices may decrease. Again, anemia lacks specificity on its own but contributes supporting data.

Alterations in skeletal muscle proteolysis raise interest in tracking associated enzymatic regulators. Research indicates serum proteasome levels increase in cachexia. Investigations continue into activators of muscle protein degradation pathways as diagnostic candidates.

Appetite-regulating hormones offer another avenue under exploration, with cachexia linked to reduced ghrelin and elevated leptin. However, wide normal ranges limit clinical utility thus far. Findings remain inconsistent across cancer types.

Ongoing efforts aim to delineate a molecular cachexia signature, capturing genomic and proteomic variations. Single markers lack sensitivity and specificity, given cachexia's heterogeneity. A panel combining inflammatory mediators, muscle catabolism indicators, metabolites, and regulators of appetite and metabolism may prove optimal for diagnosis and tracking.

In practice currently, no single laboratory test confirms or excludes cancer cachexia. Potential biomarkers require more prospective validation. A multimodal diagnostic approach combining weight trends, symptoms, physical examination, body composition data, and selective lab tests best identifies cachexia while research continues to optimize molecular diagnostics.

Keeping a close eye on the changes in hemoglobin, albumin, and common inflammatory markers can help support clinical suspicion and standard criteria for cachexia. Improved biomarkers promise to strengthen future cachexia diagnosis and management.

Assessment Tools and Questionnaires

In addition to clinical criteria and body composition measures, several assessment tools and questionnaires exist to help screen for and characterize cancer cachexia. These patient-reported metrics complement conventional evaluation methods.

Simple screening questions provide a convenient starting point. Asking about appetite changes, enjoyment of meals, unintended weight loss, and whether clothes fit more loosely can quickly identify signs of early cachexia for further assessment.

More comprehensive questionnaires formalize symptom screening. The anorexia/cachexia scale (A/CS) includes 12 questions on appetite, early satiety, taste, weight change, and distress regarding appetite loss. Scores correlate with cachexia severity and prognosis. The functional assessment of anorexia/cachexia therapy (FAACT) tool similarly surveys eating habits, appetite, hunger, and weight changes through 18 items.

The MD Anderson Symptom Inventory (MDASI) incorporates questions on appetite, weight loss, and muscle mass alongside assessments of fatigue, weakness, and global symptom burden. Integrating cachexia-specific concerns into broader symptom measurement facilitates a multidimensional view of its functional impacts.

Quality of life questionnaires provide insights into cachexia's effects on physical function and emotional well-being for a more patient-centered evaluation. Tools like the EORTC QLQ-C30 incorporate symptom, functional, psychological, and social domains relevant to cachexia. The FAACT quality of life instrument includes physical, functional, and emotional elements.

Physical performance status tests such as timed get-up-and-go, gait speed over short distances, grip strength, and manually assessed muscle power grade changes reflect diminished strength and endurance from cachexia. Though subjective, these tests easily screen for functional decline at the bedside.

Patient-generated subjective global assessment (PG-SGA) facilitates nutritional assessment by self-reporting recent weight changes, dietary intake, symptoms, and functional capacity. When combined with a doctor's physical exam and a review of metabolic parameters, the PG-SGA creates a comprehensive picture of a person's nutritional status that can help identify cachexia.

No single tool serves as the definitive cachexia assessment, but thoughtfully selecting patient-centered questionnaires and functional tests provides valuable insights into the psychosocial and functional impacts complementing traditional cachexia measures.

A multidimensional cachexia assessment that looks at weight changes, lab results, body composition analysis, symptom surveys, quality of life evaluations, and basic functional testing is the best way to get a full picture of this complicated syndrome that can help with diagnosis, staging, treatment, and monitoring.

Challenges in Diagnosis of Cancer Cachexia

Despite growing research, diagnosing cachexia remains challenging due to its complex multifactorial nature, evolving manifestations, and the limitations of current assessment criteria and tools. Key issues complicating diagnosis include:

1. **Lack of a universally accepted standard definition:** Consensus criteria provide a foundation but continue to be refined as evidence on optimal diagnostic cutoffs and measures emerges. Definitional variations make consolidating data across studies difficult. Updated consensus guidelines are needed to support consistency.
2. **Overlap with related syndromes:** Cachexia shares features with sarcopenia, anorexia, and malnutrition. Distinguishing primary drivers can prove difficult, especially early on. Patients may be misclassified or have overlapping etiologies missed. A multivariate approach examines interacting components.

3. **Insensitivity to weight loss:** Small losses of muscle mass may be masked by fluid shifts or remaining fat stores. Significant depletion can occur without meeting weight loss diagnostic thresholds. Measures of muscle area better characterize early wasting, but availability limits use.
4. **Fluctuations during cancer treatment:** Weight and appetite changes from anti-cancer therapies complicate identifying cachexia versus transient effects. Assessing trends before, during, and after treatment helps determine the primary drivers.
5. **Monitoring difficulties:** Frequent formal body composition assessments could be more practical. Spot checks may miss interim changes. Self-reported intake and weight are prone to recall errors or misperceptions. New remote monitoring methods are needed.
6. **Under-recognition by clinicians:** Lack of screening practices, workload constraints, and knowledge gaps result in missed opportunities for diagnosis. Training and guidelines promote

assessment where awareness lags—patient-reported metrics aid detection.

7. **Lack of definitive laboratory tests:** Reliable biomarkers still need to be discovered. Inflammation markers lack specificity. Proteolysis regulators require validation. A multi-analyte panel approach appears most promising, but optimal combinations remain undefined.

Despite these challenges, simple tools like weight trends, symptom questionnaires, quality-of-life evaluations, and physical function tests can significantly improve cachexia assessment. A multidisciplinary, longitudinal view combining clinical, patient-centered, and emerging molecular measures will continue to advance diagnosis.

Further research promises to refine definitional criteria, delineate clinical stages, and validate more precise biomarkers to support the optimal identification of this complex syndrome.

Chapter 3

Clinical Implications and Prognosis

The Impact of Cachexia on Clinical Outcomes

Cachexia profoundly impacts clinical outcomes in cancer patients, including survival duration, treatment tolerance, response to therapies, and functional performance status. The severity of cachexia correlates closely with the prognosis across tumor types.

Multiple studies demonstrate cachexia's association with reduced survival times. A 5–10% loss of pre-illness weight correlates with a markedly worse prognosis across cancer stages. Cachexia is estimated to contribute directly to up to 20% of cancer deaths. The precise mechanisms linking cachexia to mortality remain unclear but likely involve metabolic dysfunction, immunosuppression, accelerated muscle catabolism, and malnutrition.

Cachexia decreases the delivery and efficacy of cancer treatments. Depleted protein reserves compromise organ function, wound healing, and immunity, increasing toxicity risks with chemotherapy, radiation, or surgery. Medication doses often require a reduction in cachectic patients, which may negatively impact outcomes. Cachexia also promotes treatment resistance through mechanisms that are not fully understood.

Reduced muscle mass from cachexia leads to diminished functional status. Fatigue, weakness, and impaired mobility contribute to frailty. Patients struggle to complete activities of daily living and lose independence. Diminished performance status limits the ability to comply with rigorous therapy regimens.

Cachexia-related fatigue also reduces quality of life. Anorexia, dysgeusia, and early satiety degrade enjoyment of food. Emotional distress accompanies wasting. Social isolation increases as mobility declines. Symptoms cluster to diminish patient well-being profoundly.

Consequently, cachexia has been associated with decreased rates of anti-cancer treatment completion, an increased frequency of dose-limiting toxicities, a heightened risk of chemotherapy-related febrile neutropenia, and a higher incidence of hospitalization. Patients with cachexia often require increased supportive care interventions.

Cachexia must be recognized as an important prognostic factor influencing key outcomes. Routine screening for early identification, coupled with aggressive supportive care management, is warranted to help mitigate its multi-system effects. Diagnosis frequently prompts treatment modifications and greater vigilance in monitoring for complications.

However, it is important to note that cachexia reversibility varies. Intermediate outcomes can still improve with appropriate interventions, even in advanced cachexia. Nutritional therapy, physical activity, appetite stimulants, and medications targeting catabolic pathways may help stabilize Cachexia, offering quality-of-life benefits. Patients should not be undertreated simply due to a cachexia diagnosis when supportive options exist.

Identification of cachexia provides key prognostic information that should prompt heightened monitoring, evaluation of the need for treatment modifications, increased supportive care, and additional counseling around expectations. However, conscientious cachexia-focused management still aims to optimize quality of life and function. Ongoing research also seeks to develop interventions that could counteract the adverse effects of cachexia on cancer prognosis.

Cachexia Staging and Expected Prognosis

The severity of Cachexia is closely associated with the prognosis. Patients with early-stage cachexia generally have more favorable outcomes compared to those with refractory, end-stage wasting. Formal cachexia staging supports conveying a likely prognosis and guides management goals.

In early pre-cachexia, patients demonstrate subtle metabolic changes and mild appetite loss without significant weight or muscle loss. The prognosis remains driven predominantly by the cancer stage at this point. Pre-cachexia signals increased risks for progression to overt cachexia, requiring proactive monitoring and prevention efforts.

As measured weight loss exceeds 5%, patients meet consensus criteria for cachexia. Even minor cachexia of 5–10% weight loss worsens the prognosis across tumor types. Median survival durations are measured in months for severe versus non-cachectic patients. Supportive care focuses on stabilizing further waste to extend the quality of life.

Moderate Cachexia with weight loss of up to 15% further elevates the risks of mortality and chemotherapy toxicity. Functional deficits emerge, impacting independence. The prognosis is guarded, and cachexia management becomes central to care.

Preventing progressive wasting and maintaining energy enables continued cancer therapy.

In severe or refractory cachexia with extreme weight and muscle loss exceeding 15%, the prognosis is grave. Median survival is shortened to weeks or months. Function is profoundly impaired, and quality of life is poor. Cancer treatment often ceases. The goals shift entirely to symptom palliation and preservation of remaining function for as long as possible.

However, the prognosis remains complex. Cachexia staging provides a guide, but many factors influence outcomes. The prognosis tends to be worse when cachexia progresses rapidly if it occurs in the context of advanced or unstable cancer and when the onset is at an older age. Supportive care may still attenuate further wasting to extend and improve the quality of life.

Markers of systemic inflammation provide additional prognostic information. Elevated CRP levels portend worse outcomes, independent of cachexia stage or tumor variables. Ongoing muscle loss despite interventions indicates refractory cachexia.

The trajectory of cachexia over time also matters. Stabilization or modest improvements in weight and function with supportive care are encouraging signs, even in severe cachexia. Sustained progression despite optimal treatment portends decline. Frequent reassessment allows prognostic updates.

Cachexia severity based on the degree of involuntary weight and muscle loss provides a useful framework for conveying the expected prognosis and guiding care goals tailored to where a patient falls on the spectrum. However, attention to the full clinical picture allows personalized prognostic estimates to inform compassionate patient-centered care.

Effects of Cancer Cachexia on Quality of Life

Cancer Cachexia profoundly diminishes the quality of life through diverse physical, emotional, and psychosocial effects. Understanding and addressing quality-of-life impacts is key to patient-centered cachexia care.

Physically, reduced muscle strength from cachexia impairs mobility, balance, and the ability to perform activities of daily living—fatigue and exhaustion increase. Diminished exercise tolerance limits independence. Household chores, shopping, and personal care become challenging. Frailty elevates the risk of falls and fractures.

Respiratory function declines as intercostal and accessory muscles waste. Patients experience dyspnea at rest or with minimal exertion. A weak cough impedes airway clearance. Ventilatory capacity decreases, and lung infections become more common.

Cardiac atrophy and low albumin from Cachexia reduce circulatory reserve. Orthostatic hypotension may occur. Patients describe racing heart rates or palpitations with minor activities due to impaired cardiac response. Reduced perfusion contributes to fatigue.

Appetite loss, taste changes, and early satiety seriously degrade the enjoyment of eating. Nutrient depletion exacerbates weakness and low energy.

Diarrhea or constipation resulting from reduced intake further diminishes the quality of life.

Emotionally, increased dependence on others for basic needs leads to frustration and distress. Helplessness and isolation arise as Cachexia encroaches on lifestyle. The visible manifestations of wasting provoke anxiety and worsen depression.

Body image concerns emerge as weight and muscle loss alter the physique. Changes feel unfamiliar and reinforce declining health, especially for patients who valued fitness previously. Some withdraw socially to avoid distress or awkward comments about weight loss.

Information gaps about cachexia exacerbate uncertainty and fear over the control of symptoms or trajectory. Open discussions explaining expected outcomes allow psychological preparation and adaptation. Counseling provides emotional support.

Medication side effects, frequent lab draws, and medical appointments all contribute to feelings of burden. Financial stress arises due to the costs of supportive care, home modifications, and loss of work capacity. Roles in families and society are disrupted.

Addressing quality of life allows tailored interventions that reflect patient priorities. Appetite stimulants, nutritional supplements, mobility aids, breathing techniques, counseling, and social work support aim to preserve functioning and well-being. The key is open communication about cachexia's holistic impacts.

In advanced illness, quality of life becomes central. Careful attention to Cachexia's physical and psychosocial effects remains integral, even when reversing wasting proves difficult. Integrated supportive care continues to promote comfort, dignity, and patient autonomy.

Health Economics and Healthcare Utilization Related to Cancer Cachexia

Cachexia contributes significantly to the economic burden of cancer care through increased healthcare costs, service utilization, hospitalizations, and loss of work productivity. Quantifying these impacts sheds light on the importance of cachexia prevention and management.

Direct medical costs arise from medications, nutritional supplements, clinical appointments, testing, and increased supportive care needs driven by Cachexia. Indirect costs include diminished work capacity or job loss for patients and caregivers and transportation for frequent care.

Hospital admissions increase markedly in Cachexia, given associated immune dysfunction, higher chemotherapy toxicity risks, infections, and declining self-care ability. Costs per admission average thousands of dollars, accumulating quickly. ICU stays are also more likely, adding substantially to expenses.

Outpatient services are utilized more frequently in efforts to provide cachexia symptom management and preserve quality of life. This includes more regular labwork, clinic visits, procedures, imaging, and consultations with nutrition, rehabilitation, or mental health specialists. Home health services are often employed.

Cachexia care medications further increase costs. Appetite stimulants, nutritional supplements, anti-inflammatory agents, and therapies targeting muscle catabolism can cost hundreds of dollars monthly, especially newer biologics. Supportive injectables like erythropoietin for anemia also increase.

Indirect costs arise from lost work days for patients, caregivers, and family members. Job loss or early retirement due to disability cuts income while increasing public aid needs. Reduced performance and absenteeism affect employers.

Analysis suggests that cachexia at least doubles total direct cancer care costs over the disease course. Costs correlate with cachexia severity, becoming up to 8-fold higher in refractory cachexia. Expenses escalate rapidly as cachexia advances, given the impacts on hospitalization, loss of independence, and supportive care needs.

Upfront investment in screening, early intervention, and cachexia-focused care aims to improve outcomes while reducing acute resource use from delayed diagnosis and subsequent rapid deterioration. Nutritional therapy, physical rehabilitation, and targeted medications support quality of life and function to avoid crisis care.

Further research quantifying cachexia-related costs across settings and stages can support policy efforts to provide coverage for necessary supportive care that may offer overall health system savings. Quality improvement initiatives also help streamline evidence-based care.

Addressing cancer cachexia remains challenging but critical to improving patient prognosis, function, and experiences. Ensuring access to appropriate multimodal care while minimizing wasted spending has mutual benefits for quality of life and the effective use of limited resources. Awareness of Cachexia's economic impacts will continue to drive improvements.

Ethical Considerations Related to Cancer Cachexia

Cachexia raises several ethical issues for clinicians and families regarding goals of care, palliative integration, and advanced planning as cachexia-related illness burdens accumulate. Sensitive communication promotes shared understanding and patient-centered decision-making.

At diagnosis, discussions should explore prognosis, expected cachexia trajectories, and potential impacts on quality of life. This grounds later conversations when cachexia arises or advances. Patients can preemptively express wishes and values to guide care as function declines.

As cachexia develops, discussions should revisit the prognosis in light of recent changes. Reviewing realistic outcomes gives patients the autonomy to set priorities and make informed choices. Clinicians have an ethical duty to convey truths sensitively but candidly.

The focus should shift from cachexia cure to palliation as cachexia enters refractory stages. Priorities align with comfort, dignity, and the quality of remaining alive rather than prolongation. Palliative interventions integrate into active cachexia management early on.

Nutrition and hydration require nuanced discussions in advanced cachexia. Patients may experience eating as a chore rather than a pleasure. A lower caloric intake may provide comfort by decreasing nausea or diarrhea. The cachexia trajectory, despite interventions, should guide choices.

Similarly, the benefits and burdens of blood transfusions for cachexia-associated anemia merit consideration. Quality of life, not lab values alone, should direct care. However, sedation from anemia may impair judging benefits versus accepting rest.

As capacity declines, reviewing surrogate decision-makers and completing advanced directives provides ethical reassurance that care aligns with patient wishes. Respect for autonomy extends into states where patients cannot articulate preferences.

Referral to palliative care specialists supports proxy decision-making consistent with patient values. Palliative teams provide added counseling and symptom management guidance as Cachexia progresses. Early integration of palliative approaches needs to be more utilized.

Of equal importance is addressing the psychosocial and spiritual distress triggered by cachexia. Social work and chaplaincy provide avenues to process grief, fear, isolation, and existential concerns that influence coping.

The ethics of cachexia care focus on transparent communication, upholding patient priorities, shared decision-making on interventions, early integration of palliative approaches, and holistic support for emotional suffering. Listening sensitively to concerns guides care centered on preserving quality and meaning throughout Cachexia's trajectory.

Chapter 4

Prevention and Management

The Goals of Cancer Cachexia Treatment

The overarching goals of cachexia care are to preserve the quality of life and maximize function for as long as possible through a multimodal approach. While reversing cachexia proves challenging, especially in advanced stages, focused interventions aim to:

- Attenuate further weight and muscle loss.
- Stabilize body composition.
- Optimize nutritional intake and status.
- Reduce catabolism and inflammation.
- Address anorexia and fatigue.

- Maintain strength, performance status, and independence.
- Enable continued cancer therapy.
- Ease psychosocial distress and limitations.

In early cachexia, the goal was to prevent progression from pre-cachexia to severe depletion. Nutritional optimization, physical activity, and anti-inflammatory medications slow the downward catabolic spiral. Goals focus on maintaining stamina for anti-cancer treatments, daily activities, and appetite.

In established cachexia, multimodal interventions seek to mitigate ongoing losses and achieve short-term gains in weight, muscle mass, and energy. Realistic aims are incremental improvements or plateaus rather than dramatic reversals. Supporting nutrition, anabolic approaches, social needs, and quality of life takes priority.

In refractory cachexia, cachexia management integrates fully with palliative care. The focus shifts from correcting waste toward comfort and quality time. Approaches aim to conserve remaining function while easing the burden of symptoms. Aggressive cachexia therapies often cease providing comfort-focused care.

Across stages, open discussions about cachexia goals allow for the customization of each patient's priorities. Some favor maximizing even brief life extension regardless of quality, while others focus on enjoying time without intensive therapies. Patients' choices guide shared decision-making on suitable interventions aligned with personal values.

Given cachexia's poor reversibility, expectations for treatment should be realistic but hopeful. Clinical improvements sometimes manifest through stabilization, hitting "pause" on cachexia's downward trajectory to prolong independence and participation in meaningful activities. Supportive care continues to provide benefits for those with advanced cachexia.

The overarching cachexia treatment goals are to optimize nutritional status, reduce catabolism, support function and independence, enable anti-cancer therapy, and improve quality of life. The specific needs and preferences of individual patients guide tailoring multimodal cachexia care throughout the cancer journey, focused on living life as fully as possible. Research continues to refine therapeutic strategies that balance clinical efficacy with the preservation of overall well-being.

Pharmacologic Therapies

Several classes of medications aim to counter mechanisms underlying cachexia to preserve weight, muscle mass, and function. Current drug options include corticosteroids, progestational agents, orexigenic medications, and therapies targeting inflammation and protein catabolism.

Corticosteroids such as prednisone and dexamethasone have the longest history of use in cancer cachexia. They may stimulate appetite and produce short-term gains in weight through anti-inflammatory effects and fluid retention. However, effects attenuate over weeks. Long-term use risks muscle wasting, immunosuppression, and complications.

Progestational agents like megestrol acetate more potently improve appetite and weight. They modulate neuropeptide metabolism to increase hunger signals. Small studies demonstrate modest weight increases and some improvement in quality of life, but weight gains represent mostly fat rather than muscle.

Orexigenic medications target appetite loss. Synthetic cannabinoids like dronabinol act on cannabinoid receptors to augment appetite. Clinical trials show small benefits for weight and quality of life, but side effects like somnolence limit use. Other options include mirtazapine and ghrelin analogs.

Newer biologics aim to suppress inflammation, driving cachexia. Early results with monoclonal antibodies inhibiting IL-1 and IL-6 signaling demonstrate reduced resting energy expenditure and protection of lean body mass in clinical trials. Further studies continue to define long-term efficacy.

Agents like *enobosarm* target muscle wasting through selective androgen receptor modulation to improve muscle protein synthesis. Early-phase trials have shown improved lean mass. However, impacts on strength and function remain under investigation. Concerns exist for potential abuse.

Despite increasing options, no single breakthrough medication reliably reverses cachexia. Given their complex mechanisms, multi-agent regimens targeting nutrition, anabolism, catabolism, and inflammation simultaneously provide the best outcomes but require further optimization. Appetite stimulants and anti-inflammatory medications offer accessible options to integrate with nutritional therapies. Research continues to expand the pharmacological toolkit to combat cachexia.

Nutritional Interventions

Nutritional support forms the foundation of cachexia management. Optimizing caloric intake and providing protein supplementation aims to reduce catabolism, improve weight and strength, support immunity, and enhance the quality of life. A personalized, multi-faceted nutrition approach is key.

Increasing overall caloric intake seeks to address cachexia's hypermetabolic state and high resting energy expenditure. However, supplementation must balance benefits while minimizing gastrointestinal side effects or exacerbating reduced appetite in cachexia.

Starting with smaller, more frequent meals and nutrient-dense foods enhances tolerance and intake. Calorie-dense oral supplements between meals provide extra calories and protein without filling them. Examples include juices, smoothies, and commercial shakes. Low-volume liquid nutrition helps minimize early satiety.

Where oral intake remains inadequate despite efforts, tube feeding can deliver targeted nutritional support. Enteral feeding overcomes reduced appetite while bypassing absorption issues and vomiting. Considerations include gastrostomy versus nasogastric access, bolus versus continuous regimens, and specialized formulas.

It is critical to improve protein intake to maintain lean muscle mass and wound healing. Protein requirements are elevated in cachexia to a minimum of 1.2 grams per kilogram of ideal body weight daily and up to 2 grams per kilogram in severe catabolism. Whey protein supplements offer high oral bioavailability. However, protein breakdown remains accelerated in cachexia despite intake. Strategies like increased meal frequency and pulsed high-protein delivery aim to overcome this hurdle by maximizing muscle protein synthesis between catabolic periods.

Micronutrient needs are also increased in cachexia. Supplements help correct common deficiencies of vitamins, minerals, and electrolytes depleted by cachexia. Multivitamins prevent the exacerbation of malnutrition.

Succeeding with nutrition interventions depends on appropriate timing, tailoring regimens to patient preferences and tolerances, closely monitoring intake and weight trajectories, and adjusting results-based approaches. Patients benefit from dietitian consultations for personalized plans.

Research continues on specialized oral supplements enriched with amino acids, fatty acids, and anti-inflammatory compounds to target cachexia mechanisms. Parenteral nutrition offers an alternative route when intestinal function fails but still lacks clear benefits over enteral feeding. Overall, combining counseling, oral supplements, and tube feeding as needed aims to mitigate cachexia by providing key nutrients.

The Role of Exercise and Physical Activities

Alongside nutrition, structured exercise, and daily physical activity provide critical non-pharmacologic cachexia therapies. Physical rehabilitation programs counteract muscle wasting and weakness to improve strength, function, and quality of life.

Resistance training represents a cornerstone. Lifting weights or resistance bands at moderate intensity stimulates muscle protein synthesis. The anabolic effects help counter proteolysis and fiber atrophy in cachexia. Even in advanced cancer, resistance training is safe and improves strength. The impact depends on frequency, volume, and load progression under guidance.

Aerobic activity such as walking, cycling, or swimming complements resistance exercise. Low- to moderate-intensity aerobic efforts augment cardiorespiratory endurance depleted by cachexia while avoiding excessive fatigue. This enhances mobility, mood, and the ability to perform daily activities. Activity should remain below the anaerobic threshold.

Physical therapy also focuses on maintaining range of motion through stretching to prevent contractures as mobility declines. Compression garments can reduce limb edema that develops as protein depletion impairs oncotic pressure. Assistive walking devices preserve safe mobility with weakness.

Timing activity around meals aims to optimize nutrition delivery to muscles when anabolic sensitivity is highest. Brief bouts of resistance activity during the postprandial phase may improve nutrient partitioning in muscle. Small, frequent meals coupled with periods of movement provide cyclic nutrition and exercise mini-bursts throughout the day.

In advanced cachexia, lighter resistance bands and supported partial range movements allow continued activity as tolerance decreases. Chair exercises promote mobility. Efforts shift to maintaining existing strength and range of motion for as long as possible for basic function. Preventing prolonged immobility also wards off related complications.

Patient engagement increases by tailoring activities to individual interests and functional levels. Realistic goal-setting focuses on specific achievements like walking a certain distance or completing personal care tasks to foster confidence. Therapeutic exercise should not worsen fatigue or reduce enjoyment of life.

In addition to structured rehabilitation, counseling promotes baseline physical activity and minimizes sedentary time. Light household chores, leisure walking, and recreation support general fitness as much as possible. However, overexertion that outpaces capacity requires avoidance as well.

Research continues to optimize rehabilitative approaches. Multimodal exercise programs best counteract cachexia when combined with nutrition and pharmacology. Physical activity provides benefits across the cachexia spectrum when individually tailored and guided.

Behavioral Strategies

Incorporating behavioral strategies enhances cachexia care by addressing anxiety, low mood, coping, disordered sleep, and maladaptive thoughts that often accompany chronic illness. Psychosocial support is key to optimizing well-being.

Counseling provides a safe outlet for patients to discuss fears, frustrations, grief, and other emotions amplified by cachexia. Therapists help reframe negative mindsets, overcome demoralization, and guide realistic goal-setting. Counseling aims to help patients process and adapt to the losses that cachexia imposes.

Support groups led by social workers or peers also provide communal understanding for those experiencing cachexia. Sharing stories, advice, and encouragement creates connection and reminds patients that they are not alone in their struggle. This counseling fosters hope.

Cognitive behavioral therapy (CBT) techniques aid patients in identifying and modifying dysfunctional thought patterns like catastrophizing or helplessness that lead to distress and hinder coping. CBT exercises teach healthy mental approaches to life's challenges. These skills support resilience.

Progressive muscle relaxation, meditation, guided imagery, and hypnosis help patients cultivate calming responses to anxiety and pain amplified by cachexia. Deep breathing with visualization provides in-the-moment relaxation. Yoga and mindfulness practices induce mental stillness.

For cachexia-associated insomnia, behavioral sleep hygiene methods facilitate better rest. Keeping a consistent routine, limiting naps, creating an ideal sleep environment, and curtailing evening stimulation improve sleep quality and daytime alertness. Relaxation aids in reducing nighttime drowsiness.

Appetite-focused behavioral plans use scheduled meal times, food cues, new recipes, and family meals to encourage oral intake and counter poor hunger signals in cachexia. Patients keep intake logs to identify eating barriers and motivators to optimize nutrition.

Guidance on energy conservation techniques helps patients adapt activities to match the reduced stamina imposed by cachexia. Clustering tasks, using mobility aids, and spacing rest periods preserve independence and avoid exhaustion. Realistic pacing fosters function.

While not curative, thoughtful behavioral interventions meaningfully complement cachexia medical care. They empower patients with coping skills, targeted strategies, and peer support to nourish quality living. Integrative cachexia care should incorporate psychosocial counseling, education, and training.

Multimodal Approaches

Given the complex metabolic derangements underlying cancer cachexia, management strategies that deploy multiple complementary modalities in parallel produce superior outcomes compared to isolated approaches. Thoughtfully developed multimodal care plans are considered optimal practice.

Combining medications provides synergistic benefits. Appetite stimulants address reduced oral intake, anabolic agents aim to improve muscle protein synthesis, and anti-inflammatory drugs suppress catabolism. Multidrug regimens treat diverse cachexia drivers simultaneously.

Adding structured nutrition support to pharmaceutical management enhances the impact. Timed, high-calorie, high-protein supplements augment oral intake alongside medications, improving appetite and anabolism. Nutrition provides the fuel and building blocks of medications, which are partitioned most efficiently.

Integrating tailored exercise training complements nutraceutical efforts to maintain muscle mass. The anabolic effects of resistance training coupled with aerobic activity's enhancement of resting metabolism add physical rehabilitation to biochemical therapy. Movements oppose immobility.

Behavioral strategies foster the success of other interventions by promoting intake, activity, functional independence, and mental resilience. Counseling improves compliance with regimen changes. Coping skills empower patients to overcome cachexia-related limitations.

Care coordination optimizes the benefits of each modality. Oncology, palliative care, nutrition services, physical and occupational therapy, and psychosocial providers synchronize input. Consensus on the cachexia stage and trajectory allows patient-centered goal-setting. Defined roles prevent duplication.

The sequence of implementing modalities matters. Nutrition and exercise may first stabilize overt catabolism. Medications can then accelerate anabolism as intake improves. Behavioral therapy facilitates adaptation and adherence. Each modality prepares the patient to gain maximal benefit from the next.

Multimodal therapy also mitigates the adverse effects of single interventions when applied judiciously. Lower medication doses may suffice when combined with nutrition and training. Adverse drug effects are avoided.

While evidence continues to evolve, current best practice endorses an individualized multidisciplinary combination of medical, nutritional, physical, and psychosocial interventions tailored to the cachexia stage and patient goals. Multimodal therapy reflects cachexia's complex pathophysiology, requiring an integrated, systematic cachexia-focused treatment approach.

Evaluating the Treatment Response

Ongoing monitoring helps clinicians determine if cachexia interventions provide benefits versus the continued progression of wasting that may warrant modifications to the treatment approach. Important parameters to track include:

1. **Weight trends:** Stabilization or small gains suggest effective attenuation of catabolism. Continued weight loss flags inadequate control of cachexia progression. Frequent weight checks are key.
2. **Muscle mass:** Imaging or strength assessments periodically evaluate muscle area

since fluid shifts can confound weight. Improving or maintaining mass indicates successful muscle-focused therapy.

3. **Appetite and oral intake:** Patient-reported hunger, meal completion, and calorie counts assess appetite stimulation and nutrition efficacy.

4. **Inflammatory markers:** Downward trending CRP, IL-6, or ESR suggests anti-inflammatory efforts are reducing catabolic drive. Levels rising above baseline indicate persisting inflammation.

5. **Hemoglobin/albumin:** Improving levels confirm reduced inflammation and adequate protein-calorie intake. Declines signal persistent cytokine activity or ineffective nutrition support.

6. **Physical function:** The ability to complete standardized tasks like timed walks, grip tests, and chair stands reflects retained or recovered strength from therapy. Inability denotes progression.

7. **Patient-reported outcomes:** Appetite, quality of life, and symptom scores on validated scales add to the patient's perspective on multidimensional cachexia control and functional impacts.
8. **Cancer treatment tolerance:** Completing prescribed anti-cancer therapy indicates preserved strength and nutritional status to withstand regimens. Interruptions or dose reductions raise concern.

The optimal frequency of reassessment depends on the cachexia stage, treatment intensity, and goals of care. During active therapy changes, weekly checks may help gauge the effects. For maintenance, monthly suffices. A new symptom onset should prompt evaluation.

Based on the results, careful trend analysis and program modification improve the chances of stabilizing cachexia over time. Realistic treatment goals are stabilization or slowed progression rather than radical reversal in advanced cachexia. Preventing rapid deterioration marks success.

Challenges in the Treatment of Cancer Cachexia

Despite expanding therapeutic options, effectively managing cachexia in clinical practice remains challenged by issues around adherence, adverse effects, access to services, and limitations in reversing advanced wasting.

Poor adherence to lifestyle changes and medication regimens frequently limits cachexia therapy. Appetitive stimulation and calories alone cannot overcome catabolism without patient engagement. Taste fatigue, forgetfulness, costs, depression, or counseling gaps reduce compliance. Motivational support and clear education promote adherence.

Managing adverse effects like diarrhea, early satiety, fatigue, and somnolence improves the tolerance of cachexia regimens. Adverse effects exacerbate disability and emotional distress. Individualized dose adjustments, sequencing, or alternative therapies help retain quality of life while treating cachexia.

Access barriers related to insurance, affordability, transportation, and the availability of specialist services impede optimal cachexia care for some patients. Nutritional supplements, activity programs, injectable medications, and home support may not be covered or have high co-pays. Advocacy helps address disparities.

Refractory cachexia with rapid depletion responds poorly to the available interventions. Patients with uncontrolled cancer progression, despite systemic therapy, have limited reversibility of weight loss and muscle wasting. In advanced illnesses, cachexia therapies focus on comfort and quality of life.

Restrictions around the use of appetite stimulants, growth factors, and anabolic agents in cancer complicate pharmacologic options. Concerns exist about promoting tumor growth or recurrence. Supportive care aims to counteract cachexia while ensuring oncologic safety.

Sustaining motivation for intensive cachexia treatment regimens is challenging as the illness progresses and treatment fatigue sets in. Ongoing counseling helps reframe goals from reversing cachexia to stabilization and maximizing quality time.

Despite progress made in cachexia therapies, further research remains vital to expand management tools, optimize treatment algorithms, improve access, and provide patients and families with the education and support needed to persevere with cachexia care. A multidisciplinary approach addresses complex nutritional, medical, functional, and psychosocial needs in cachexia.

Chapter 5

Special Considerations

Cachexia in Association With Other Chronic Conditions Beyond Cancer

While most commonly discussed in malignancies, cachexia syndromes are associated with a range of chronic conditions beyond cancer, including COPD, heart failure, rheumatoid arthritis, and other respiratory, cardiac, renal, and inflammatory diseases.

In COPD, skeletal muscle wasting and unintended weight loss are predictive of mortality, independent of lung function. Approximately 20–40% of COPD patients become cachectic, accelerating disability. Reduced appetite, increased muscle protein breakdown, and inflammation drive COPD cachexia. Systemic corticosteroids used in treatment may also exacerbate muscle wasting over time.

Cachexia is a key predictor of mortality in chronic heart failure, linked to sympathetic over-activation, inflammation, and impaired anabolic signaling. Mechanisms overlap cancer cachexia but also involve direct cardiac cachexia factors that induce skeletal muscle loss. Wasting predicts poor functional status. About 10–15% of CHF patients develop cardiac cachexia.

Rheumatoid cachexia describes muscle loss and inflammation associated with rheumatoid arthritis (RA) that leads to weakness and joint damage exacerbations. RA cachexia seems driven by proinflammatory cytokines, including TNF-alpha. The prevalence approaches 2/3 of RA patients by disease duration of 10+ years. There are several distinctions in cachexia associated with other conditions compared to cancer cachexia:

- Non-cancer cachexia often relates more to primary inflammation than tumor byproducts.

- Weight loss may be less prominent, while muscle loss dominates.
- Appetite is often preserved rather than diminished.
- Prognostic correlations need to be clearer.

However, similar symptom-based treatments apply, including anti-inflammatories, activity/exercise, protein supplementation, and possible appetite stimulants. Addressing the underlying disease process also helps attenuate cachexia.

The key is recognizing cachexia's manifestations beyond cancer settings. Diagnosis is challenging if weight loss is subtle. A high index of suspicion in chronic inflammatory illness allows early nutrition and lifestyle support to counter depletion and associated outcomes. More research is needed on non-cancer cachexia to optimize management.

While cachexia is most commonly characterized by malignancy, providers in other specialties managing chronic conditions must recognize analogous wasting syndromes that significantly impact patient prognosis and quality of life. Early intervention stands to improve a range of outcomes in COPD, heart failure, RA, and other patient populations vulnerable to cachexia.

Unique Factors Related to Cachexia in Certain Types of Cancer

While cancer cachexia has common features across malignancies, some distinctions emerge in cachexia manifestations and recommended management strategies based on the primary tumor site, particularly in pancreatic, gastrointestinal, and lung cancers.

Pancreatic cancer is associated with the highest prevalence and most rapid, severe progression of cachexia. Mechanisms center around pancreatic secretion of lipid mobilizing factor and proteolysis-inducing factor, driving dramatic depletion of adipose and muscle tissues. Pancreatic cancer cachexia is exceptionally refractory to interventions.

In *gastrointestinal cancers*, malnutrition, and diarrhea from mechanical obstruction or malabsorption directly exacerbate cachexia. Requirements for higher protein and caloric supplementation are combined with antidiarrheals and electrolyte replacement. Enteral feeding access may be required to provide adequate nutrition.

Lung cancer induces cachexia through the release of proteases and inflammation-triggering substances. Coupled with underlying COPD that accelerates muscle wasting, lung cancer patients experience pronounced weakness and fatigue. Pulmonary cachexia impacts respiratory muscle function and oxygenation.

Head and neck cancers provoke taste changes, dysphagia, and poor dentition that undermine nutrition and oral intake. Neck radiation and surgery risks exacerbate eating challenges. Alternate nutrition delivery routes may become necessary to maintain intake.

Gynecologic and prostate cancers tend to manifest cachexia at more advanced metastatic stages but provoke profound anorexia and weakness when present. Low testosterone levels in prostate cancer patients exacerbate muscle loss.

In general, pancreatic and GI cancer patients require the most aggressive multi-modality cachexia therapy initiated early in light of rapid progression and nutrition impairment. Lung cancer patients benefit greatly from pulmonary rehabilitation tailored to their fragility. Nutritional supplements with anti-inflammatory elements show particular promise for prostate and gynecologic malignancies.

All cancer patients deserve prompt screening and cachexia-focused care, identifiable risk factors can guide heightened vigilance and proactive interventions in populations with the highest cachexia prevalence and poorest prognosis. Oncology providers should remain alert to the patterns of cachexia associated with different tumor types to optimize targeted supportive care.

Important Considerations for Cancer Cachexia Prevention and Management Across the Cancer Trajectory

Cancer cachexia manifests variably throughout the disease course. Tailoring cachexia care requires consideration of the unique priorities and opportunities at diagnosis, during treatments, and in advanced stages.

At diagnosis, screening for early cachexia signs, including weight trends, appetite, inflammatory markers, and muscle mass, establishes a baseline to guide future monitoring. Education on nutrition and activity aims to prevent the onset of cachexia. Treatment plans optimize cancer control to mitigate cachexia triggers.

During active anti-cancer therapy, cachexia risk relates to treatment effects and fluctuating inflammation. Frequent monitoring helps distinguish cachexia progression from temporary symptoms like nausea. Adjustments are made to maintain nutrition, activity, and strength through planned regimens.

Prevention of rapid cycling catabolism, or anabolism, optimizes results.

In advanced or end-stage illness, the focus shifts to relieving refractory cachexia's symptoms and preserving quality of life. Medications aim to reduce catabolism and inflammation while boosting appetite and energy. Light exercise maintains function. Goals center on comfort, meaningful time, and palliative support.

At diagnosis:
- Assess the baseline status.
- Optimize cancer treatment.
- Educate on prevention.
- Encourage nutrition and activity.

During treatment:
- Monitor fluctuations closely.
- Manage side effects.
- Adjust interventions to retain strength.
- Prevent cyclical wasting patterns.

In advanced illness:
- Focus on the quality of life.

- Palliative refractory symptoms
- Reduce catabolic medications if possible.
- Provide comfort-focused care.

The cancer stage and treatment status guide the intensity, priorities, and purpose of cachexia management. Optimizing cancer control remains paramount, while cachexia interventions support patients through predictable challenges at each phase.

With a longitudinal, dynamic view, clinicians are best positioned to anticipate cachexia risks, detect worsening trends, modify interventions, and align care with evolving needs over the cancer journey. Adaptable, patient-centered cachexia care maximizes quality of life during cancer survivorship or into advanced illness.

Special Considerations for Cancer Cachexia Prevention and Management In Vulnerable Populations

Cancer cachexia deserves attention in all patients, certain populations face distinct risks or barriers warranting focused screening and management approaches. Key groups requiring tailored cachexia care include the elderly, minorities, and those with low socioeconomic status.

Advanced age predisposes to cachexia through existing sarcopenia, reduced reserves, higher inflammation, and comorbidities. Screening starts in the early stages before severe depletion. Exercise combats weakness; nutrition addresses higher protein needs. Counseling provides coping skills and caregiver support.

However, avoiding overtreatment is also important for the elderly. Care aligns with the goals of function and independence over high-burden interventions. Conservative cachexia management aims to preserve quality of life.

Among minorities, health literacy gaps may limit awareness of cachexia prevention and treatment options. Linguistically and culturally appropriate education empowers self-advocacy and adherence. Attention to mistrust promotes engagement with cachexia-focused care.

Social determinants of health increase cachexia susceptibility and reduce care access in low socioeconomic status groups. Nutritional gaps exacerbate pre-cachexia, and income barriers block therapies. Multidisciplinary care coordination helps overcome limitations through community health workers and policy changes.

In all populations, patient-centered communication remains key. Goals of care discussions elicit individual priorities to guide cachexia management. Age, culture, literacy, language, family structure, finances, and values help customize appropriate cachexia care.

Special attention to cachexia detection and innovative management in vulnerable groups aims to address disparities and unique needs.

Elderly:
- Focus on safety and function.
- Simplify regimen complexity.
- Support shared decision-making.

Minorities:
- Provide culturally appropriate education.

- Build trust and address barriers.
- Partner with community resources.

Low SES:
- Coordinate supportive services.
- Advocate for care access.
- Emphasize feasible lifestyle steps.

A compassionate, tailored approach allows for cachexia prevention and palliation for all. Oncology providers are critical in identifying at-risk populations and enabling equitable access to supportive care through communication, advocacy, and innovation.

Cachexia Care Considerations in Palliative and End-of-Life Settings

Cachexia management takes on a different, more comfort-focused purpose during cancer palliation compared to earlier treatment stages. Key priorities include advancing directives, nutrition, energy, weakness, dyspnea, and family support.

Establishing the patient's wishes and limits on interventions guides aligned care as cachexia progresses. Discussions focus on priorities, unacceptable burdens, and expectations. Realistic choices require sensitive conveyance. Surrogate decision-makers should identify whether capacity should decline.

As appetite wanes, emphasizing favorite, high-calorie foods provides enjoyment over forced intake—assistive devices and positioning aid in eating. Providing companionship at meals supports social comfort. Ultimately, accepting reduced oral intake prevents distress.

Similarly, nutrition goals shift from reversing depletion to providing pleasure and energy and minimizing hunger. Light, frequent meals, milkshakes, smoothies, and nutrient-dense snacks suffice. If indicated based on goals, tube or intravenous feeding relieves hunger while avoiding satiety.

Managing fatigue aims to conserve energy for patient priorities. Planning focused activities with scheduled rest, obtaining equipment to preserve independence, and allowing others to take over burdensome tasks are effective strategies.

Opioids, anticholinergics, anxiolytics, and stimulants may provide pharmacologic energy as needed symptomatically. Oxygen supports dyspnea. Steroids continue to counter inflammation while avoiding long-term risks.

Family counseling equips loved ones to cope with cachexia's impacts at life's end—social work and spiritual support aid in processing anticipatory grief. The interdisciplinary team addresses psychosocial distress amplified by cachexia.

Cachexia care during palliation and end-of-life requires a supportive, comfort-focused approach. Aligning management with patient wishes and dignity takes priority over cachexia reversal. Nutrition, rest, company, and medications minimize the burdens of progressive wasting while embracing quality time.

Conclusion

Cancer Cachexia is a complex and debilitating condition arising from a combination of reduced nutritional intake and metabolic abnormalities driven by malignancy and its associated inflammation. It manifests through involuntary weight loss, muscle wasting, loss of appetite, and functional impairment. Cachexia profoundly impacts quality of life, treatment outcomes, and prognosis across cancer types.

Screening patients along the cancer continuum aids early detection before severe cachexia develops. Diagnosis integrates weight trends, symptoms, physical signs, body composition data, and labs using available consensus criteria, though work is ongoing to refine definitions. Staging cachexia severity guides management.

Cachexia increases the risks of morbidity and mortality while reducing treatment tolerance and efficacy. It leads to weakness, reduced independence, hospitalizations, and heavy psychosocial tolls. Palliative efforts focus on quality of life as cachexia becomes refractory.

A multimodal approach is currently recommended, combining nutritional support, physical activity, medications, and behavioral strategies to counter diverse cachexia drivers. Treatment aims to attenuate further wasting and preserve function for as long as possible. Monitoring helps identify the need to adjust interventions.

While cachexia care remains challenging, routine screening, early intervention, prognostic awareness, and supportive therapies provide meaningful improvements in quality and duration of life. Further research continues to elucidate cachexia mechanisms and optimize treatment algorithms.

Increased provider and patient education on cachexia sparks appropriate prevention and prompt management. Advocacy is warranted for supportive care access across populations. Only through improved recognition and response to cancer cachexia can its heavy burden be effectively alleviated.

Future Lookout for Cancer Cachexia Research and Improving Clinical Care

While understanding of cancer cachexia has progressed significantly in recent decades, optimizing management remains a work in progress. Key future directions for cachexia research and clinical care include:

- Refining diagnostic criteria and staging systems through studies delineating optimal metrics, cutoffs, and combinations of factors to identify cachexia earlier and characterize progression. Molecular biomarkers and genomic profiling may strengthen future definitions.
- Elucidating inter-individual variability in cachexia progression through studying influential genetic, lifestyle, and cancer-related factors. Improved risk stratification allows for aggressive monitoring and prevention.

- Expanding pharmacological options through developing and trialing novel agents targeting appetite stimulation, muscle synthesis pathways, cytokine production, and catabolic processes. Multidrug regimens need study.
- Optimizing nutrition support strategies through research on specialized supplements, pulsed feeding approaches, and home enteral nutrition to overcome cachexia's hypermetabolism and anorexia.
- Improving physical rehabilitation guidelines for cachexia includes combinations of resistance training, aerobic exercise, and neuromuscular electrical stimulation at different disease stages. Telehealth delivery shows promise.
- Enhancing psychosocial care integration through the development of cachexia-focused cognitive-behavioral interventions and multidisciplinary care models incorporating palliative care.
- Quantifying cachexia's health and economic impacts, including direct costs and service

utilization across settings. Cost-effectiveness studies of supportive therapies guide resource allocation.

- Increased education for providers and trainees across specialties on early cachexia recognition, staging, and multimodal management principles. Educational resources for patients and families also empower appropriate care engagement.
- Advocacy initiatives to promote access to cachexia treatments include insurance coverage, tackling disparities, care coordination, and public awareness campaigns. Policy changes support multidisciplinary, cachexia-directed care.

With continued innovation in research and clinical best practices, the future outlook for preventing and managing cancer cachexia remains hopeful. Patients today already benefit from increased understanding while laying the groundwork for the next generation of progress in effectively alleviating cachexia.

Final Remarks on Improving Outcomes for Patients Experiencing Cancer Cachexia

In conclusion, while much remains to be learned about optimally diagnosing and managing cancer cachexia, significant progress has been made in recent decades to improve outcomes for affected patients. Remaining mindful of opportunities and continuing efforts in several key areas will help further advance care and alleviate suffering from this debilitating condition.

First, maintaining an appropriate level of awareness and suspicion for cachexia development across cancer treatment settings and stages supports early recognition before severe, refractory wasting sets in. No one is beyond suspicion; cachexia deserves consideration across cancer types, stages, ages, and populations. Early action is key.

Second, taking a proactive, multidisciplinary approach facilitates the rapid initiation of indicated supportive care. Oncology, palliative medicine, nutrition, rehabilitation, and psychosocial services collaborating with a shared longitudinal view of the patient enhance cachexia prevention and management integrated throughout cancer care.

Third, combinatorial therapies integrating evidence-based nutritional, pharmacological, activity-based, and behavioral interventions provide optimal results by targeting the various facets of cachexia in parallel. While research continues to refine strategies, current best practices favor multimodal care.

Fourth, open communication and shared decision-making ensure cachexia management aligns with patient priorities, values, and evolving needs along the cancer journey. Discussing prognosis, incremental goals, and benefits versus burdens allows for customized care tailored to individuals.

Fifth, routine symptom screening and cachexia staging facilitate timely adjustments to therapy based on trajectory. Frequent monitoring paired with realistic goal setting helps retain motivation and monitor impact. Adaptability is key to progressive illness.

Finally, increased advocacy for patients and families dealing with cachexia's heavy toll promotes equitable access to emerging treatments and supportive care. Policy and public efforts provide resources to enable the full spectrum of cachexia care.

While challenges persist, the outlook for mitigating cachexia's impacts continues to improve through biomedical innovations, strengthened models of care coordination, and enforcement of cachexia's place at decision-making tables. We can reduce the burden of cancer and cachexia by working together. Our patients deserve nothing less than our sustained best efforts.

Printed in Great Britain
by Amazon